ISBN 978-0-265-82977-6
PIBN 10895706

ERRORS IN MAGNETIC TESTING WITH RING SPECIMENS.

By Morton G. Lloyd.

The torus has been much in favor as a form for a specimen to be subjected to magnetic measurements, as it gives a continuous magnetic circuit of the material under test without a joint of any kind. Moreover, its shape makes it necessary to know only two of its linear dimensions in order to calculate the constants involved. The use of this form ordinarily requires the laborious winding of magnetizing and other coils on each specimen by hand, but apparatus for commercial testing is now in use which obviates this tedious process.[1]

If the coils be wound uniformly over the entire ring, the leakage of magnetic flux will be negligible, and it may be safely assumed that the flux crossing every section of the ring is the same. The distribution of this flux over the cross section is, however, not uniform, but the flux density is greater near the inner surface of the ring. In making measurements it is customary to determine the total flux by a measurement of the electromotive force developed in one of the coils, or by the throw of a ballistic galvanometer. The average value of the flux density is then found by dividing by the area of cross section; let it be B_0. The average value of the magnetizing force can be computed from the ampere-turns and dimensions; let it be H_0. The ratio $\dfrac{B_0}{H_0}$ is taken to be the permeability corresponding to the field H_0. This is only an approxima-

---

[1] J. A. Möllinger, Electrot. Zs., **22**, p. 379; 1901.
J. W. Esterline, Proc. Am. Soc. Testing Materials, **3**, p. 288; 1903.

tion, as the average permeability is not in general given by the ratio $\frac{B_o}{H_o}$, and even if it were, we have no assurance that it is the permeability of that part of the specimen which is subjected to the average magnetizing force.[2] Indeed, we must know how the permeability varies with the radius of the ring, or with H, in order to determine the discrepancy involved; and unless we have an algebraic expression connecting these quantities, it is almost impossible to compute the quantities concerned even after a preliminary measurement has given the approximate values of B corresponding to different magnetizing forces. However, if the radius a of the section is kept sufficiently small in comparison to the radius R of the ring, the variation of permeability will also be small, and may be neglected. With given dimensions of the ring, this variation will depend upon the value of the magnetic induction, as the permeability is at first an increasing function and later a decreasing function of the induction. From the known general properties of this function the variation will be greatest for inductions near the steepest part of the magnetization curve.

If the ratio $\frac{a}{R}$ is made sufficiently small, the value of the magnetizing force at the center of the section may be taken as the average value H_o, and this is an additional simplification. This again is only an approximation, since the magnetizing force at any point is $\frac{2NI}{x}$ where x is the distance from the axis of the ring, and NI is the product of current and turns. Since the average value of $\frac{I}{x}$ is not $\frac{I}{R}$ (the reciprocal of the average value of x), H_o is not the value of H at the mean radius R. The true value of H_o can be computed for a ring whose cross section is any simple geometrical figure. The formula for a circular section was derived by Kirchhoff[3] and has been used by Rowland[4] and others, and the rectangular section was early used by Stoletow. For sheet iron, where ring stampings are piled up, the rectangular section is the only one available.

---

[2] Cf. A. Stoletow, Phil. Mag., **45**, p. 40; 1873.
    G. vom Hofe, Wied. Ann., **37**, p. 482; 1889.
[3] G. Kirchhoff, Pogg. Ann. Ergbd., **5**, p. 1; 1870; Ges. Abhandlungen, p. 229.
[4] H. A. Rowland, Phil. Mag., **46**, p. 140; 1873.

The values of H_0 for circular and rectangular sections have been computed for various values of $\frac{a}{R}$ in terms of H_R, and are given in Table I, and platted in the curves of Fig. 4.

The values for a rectangular section have been computed and published by Edler,[5] but his results are not always correct. The hyperbolic logarithms used are taken to only three significant figures, with the result that the final values are *not reliable to tenths of one per cent*, but may err as much as one-fourth of one per cent.

In determining the hysteresis of a ring specimen, two methods are available. The one is to determine corresponding values of H_0 and B_0, plat a curve between them, and measure the area of this curve. This method involves the same errors as the determination of permeability. The other method is to measure the electrical energy supplied to the ring and at the same time B_0. Alternating currents are usually used for this purpose, and this requires the determination of the form factor of the induced electromotive force in order to know the maximum value attained by B_0.[6] Secondly, eddy currents are induced in the specimen, but if not large can be approximately determined and separated by measurements at two frequencies. Moreover, on account of the nonuniform distribution of the flux in the section of the ring, the power expended is not the same as it would be if the distribution were uniform. This is owing to the fact that the energy per cycle is not proportional to the flux density B, but to some power of it, approximately B^2 for eddy currents and $B^{1.6}$ for hysteresis. Here, again, the average value of $B^{1.6}$ is not the same as the 1.6th power of the average B.

It can easily be shown that where equal volumes of material are traversed by the fluxes of different density, as in a straight bar, the loss due to nonuniform distribution must be greater than the loss with uniform distribution.[7] In the case of a ring, however, the denser flux follows a shorter path, and it has been shown by Richter[8] that if the energy is proportional to a power of the

---

[5] R. Edler, Mitteil. Techn. Gewerbe-Museums, Vienna, **16**, p. 67; 1906. Science Abstracts, 9 B, p. 158; 1906.

[6] Cf. Lloyd and Fisher, this Bulletin, **4**, p. 469; 1908: or Robinson and Holz, Gen. Electric Rev., **10**, p. 236; 1908.

[7] Cf. B. Soschinski, Electrot. Zs. **24**, p. 292; 1903.

[8] R. Richter, Electrot. Zs. **24**, p. 710; 1903.

magnetic induction between 1 and 2, the loss is less with the actual distribution than it would be with a uniform distribution. This is due to the fact that the denser flux is established near the inner surface of the ring and traverses a smaller volume than the flux near the outer circumference, and the greater loss per unit volume where the flux is dense is more than counterbalanced by the reduced volume subjected to this loss.

Consequently, in most cases of closed magnetic circuits, as in the transformer, the iron loss is less than it would be with uniform distribution of the same total flux. The dimensions are usually such as to involve very great variations in the flux density, and since the distribution of flux varies with the amount of flux (owing to varying permeability) no valid deductions as to the properties of the iron can be made from experiments in which the voltage applied to such an apparatus is varied. Apparatus of this kind may be used to determine the relative quality of different specimens of iron under the same conditions, but no reliance can be placed upon absolute values obtained under such conditions, nor upon variation of loss with B.

If we were dealing with a medium of constant permeability we could calculate the effect of nonuniformity of distribution. With iron, however, the distribution changes with each change in B_0, so that a general solution of the problem is not practicable, but particular cases may be worked out where the variation of permeability with induction is known.[9] It is well to remember that for low inductions the nonuniformity is greater, while with high values of the induction it is less than for constant permeability.

This is illustrated in Fig. 1, where the actual variation of B across the section of a certain ring is given for three different values of the magnetizing force. This ring had inner and outer diameters of 6.9 and 8.9 cm, respectively. Consequently $\frac{a}{R} = .1282$. When the value of B at the mean diameter is 2000 gausses the extreme values are 1350 and 2900, a range of 73 per cent of the mean. With 7800 gausses in the middle, the extreme values are 6700 and 8700, a range of 26 per cent. With 15,000 in the middle, the extremes are 14,500 and 15,350, a range of less than 6 per cent.

---

[9] See Richter, loc. cit.

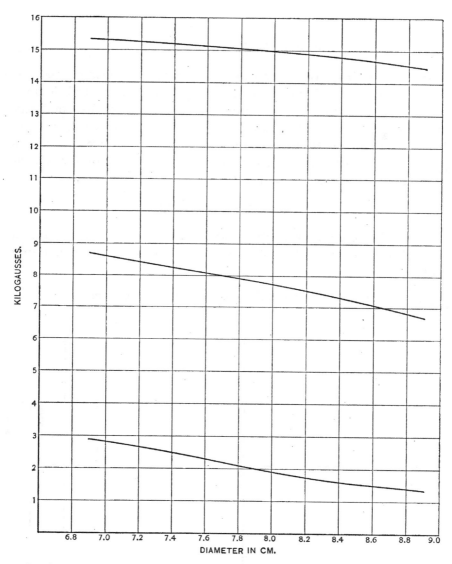

Fig. 1.—*Showing Variation in Flux Density with Distance from Axis of Ring, for Three Different Values of Magnetizing Current.*

In what follows the permeability is assumed constant, and for given dimensions of the ring the ratio of the theoretical loss for uniform distribution to the actual loss is computed, the total flux remaining constant. The solution for rings of rectangular section has been given by Richter in the paper already referred to, but is given here for the sake of completeness. The solution for a ring of circular section is believed to be new.

MAGNETIZING FORCE.

Ring of Rectangular Section.—Fig. 2 shows a section through the axis of the ring. Let NI = total current-turns of magnetizing coil uniformly distributed.

$$p=\frac{a}{R}=\text{ratio of radial width to mean diameter of ring.}$$

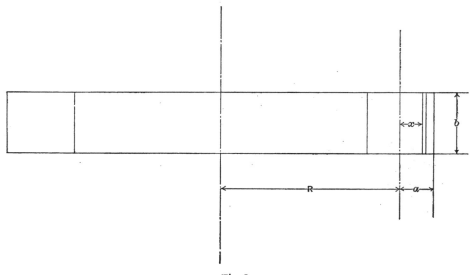

Fig. 2.

The flux in an elementary ring of thickness dx and of unit permeability is

$$d\Phi=\frac{4\pi NI}{\dfrac{2\pi(R+x)}{bdx}}=\frac{2bNIdx}{R+x}$$

$$\Phi = 2bNI \int_{-a}^{a} \frac{dx}{R+x} = 2bNI \log \frac{R+a}{R-a}$$

$$= 2NIb \log \frac{1+p}{1-p}$$

The average magnetizing force is

$$H_o = \frac{\Phi}{2ab} = \frac{NI}{a} \log \frac{1+p}{1-p}$$

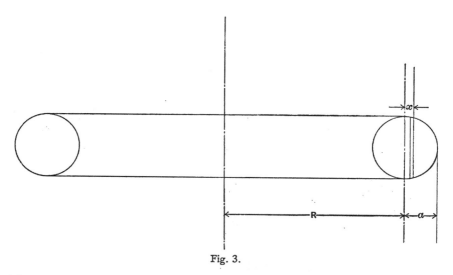

Fig. 3.

The magnetizing force at the center of the section is

$$H_R = \frac{4\pi NI}{2\pi R} = \frac{2NI}{R}$$

and

$$\frac{H_o}{H_R} = \frac{1}{2p} \log \frac{1+p}{1-p}$$

The values of $\frac{H_o}{H_R}$ are given in Table I and platted in Fig. 4 for values of p from $\frac{1}{2}$ to $\frac{1}{19}$.

Ring of Circular Section.—Fig. 3 shows section through axis of ring.

$$d\Phi = \frac{4\pi NI}{\dfrac{2\pi(R+x)}{2ydx}} = \frac{4NI}{R+x}\sqrt{a^2-x^2}\,dx$$

$$\Phi = 4NI \int_{-a}^{a} \frac{\sqrt{a^2-x^2}}{R+x}\,dx = 4\pi NI\left[R - \sqrt{R^2-a^2}\right]$$

$$H_0 = \frac{\Phi}{\pi a^2}$$

$$H_R = \frac{4\pi NI}{2\pi R} = \frac{2NI}{R}$$

$$\frac{H_0}{H_R} = \frac{2R}{a^2}\left[R - \sqrt{R^2-a^2}\right] = \frac{2}{p^2}\left[1 - \sqrt{1-p^2}\right]$$

$$= 1 + \frac{1}{4}p^2 + \frac{1}{8}p^4 + \frac{5}{64}p^6 + \frac{7}{128}p^8 + \cdots \cdots$$

TABLE I.

Ratio of the Average Value of H to the Value at the Mean Radius in Rings of Rectangular and Circular Section.

	$\dfrac{H_0}{H_R}$	
	Rectangular	Circular
$\frac{1}{2}$	1.0986	1.0718
$\frac{1}{3}$	1.0397	1.0294
$\frac{1}{4}$	1.0216	1.01625
$\frac{1}{5}$	1.0137	1.0102
$\frac{1}{6}$	1.0094	1.0070
$\frac{1}{7}$	1.0069	1.00515
$\frac{1}{8}$	1.0052	1.0040
$\frac{1}{16}$	1.0033	1.0025
$\frac{1}{19}$	1.0009	1.0007

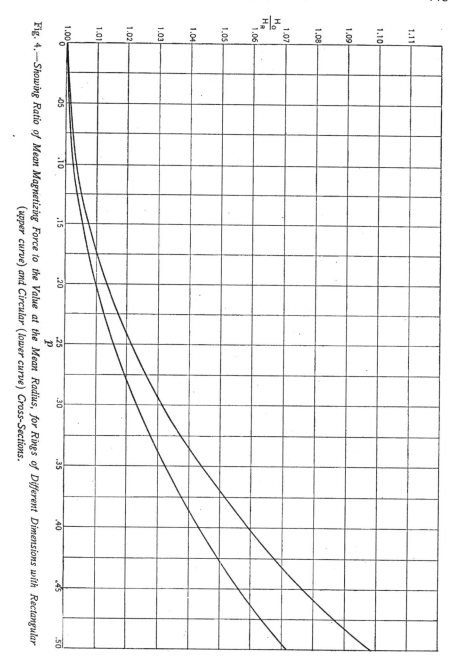

Fig. 4.—*Showing Ratio of Mean Magnetizing Force to the Value at the Mean Radius, for Rings of Different Dimensions with Rectangular (upper curve) and Circular (lower curve) Cross-Sections.*

TABLE II.

Limiting Values of p and $\dfrac{1}{p}$ for Given Allowable Error in Using H at Mean Radius for Mean H.

Per Cent Error	Rectangular		Circular	
	p	$\dfrac{1}{p}$	p	$\dfrac{1}{p}$
0.1	0.055	18.2	0.067	14.9
0.2	.077	13.0	.090	11.1
0.5	.120	8.3	.138	7.2
1.0	.172	5.8	.200	5.0
2.0	.240	4.2	.277	3.6

These values are given in Table I and plotted in Fig. 4. It is to be noticed that the values of H_0 differ less from H_R when the section is circular than when it is rectangular, for the same values of p. With a given area of section, however, it is easy to make p small for the rectangle by increasing the height b and decreasing a. Table II shows the limiting values of p in order to keep the error within assigned limits in using H_R in place of H_0.

ENERGY LOSSES.

Rectangular Section.—We assume constant permeability μ, and energy proportional to the mth power of the flux density. Let η be the energy per unit volume for unit flux density.

$$H_x = \frac{4\pi NI}{2\pi(R+x)}$$

The energy expended in an elementary ring per cycle is

$$dW = \eta \left[\frac{4\pi NI\mu}{2\pi(R+x)}\right]^m 2\pi(R+x)b\,dx$$

and in the entire ring it is

$$W = \eta(2\mu NI)^m 2\pi b \int_{-a}^{a} \frac{dx}{(R+x)^{m-1}}$$

$$= 2\pi b\eta(2\mu NI)^m \frac{(R+a)^{2-m} - (R-a)^{2-m}}{2-m}$$

unless $m = 2$. If $m = 2$

$$W = 2\pi b\eta(2\mu NI)^2 \log \frac{R+a}{R-a}$$

For uniform distribution the flux density is everywhere

$$\mu H_0 = \mu \frac{NI}{a} \log \frac{1+p}{1-p}$$

and

$$W_0 = \eta \left(\mu \frac{NI}{a} \log \frac{1+p}{1-p} \right)^m \int_{-a}^{a} 2\pi (R+x)b\,dx$$

$$= \eta \left(\mu \frac{NI}{a} \log \frac{1+p}{1-p} \right)^m 2\pi b \, 2aR$$

$$\frac{W_0}{W} = (2p)^{1-m} \left(\log \frac{1+p}{1-p} \right)^m \frac{2-m}{(1+p)^{2-m} - (1-p)^{2-m}}$$

If $m = 2$

$$\frac{W_0}{W} = \frac{1}{2p} \log \frac{1+p}{1-p}$$

The latter value is also obtained if $m = 1$ and is the same as $\dfrac{H_0}{H_R}$.

For intermediate values, $\dfrac{W_0}{W}$ is greater, reaching a maximum for a value of m equal to $\dfrac{3}{2}$, as is shown in Fig. 5, where the values of $\dfrac{W_0}{W}$ for $p=0.1282$ have been platted for varying values of m.

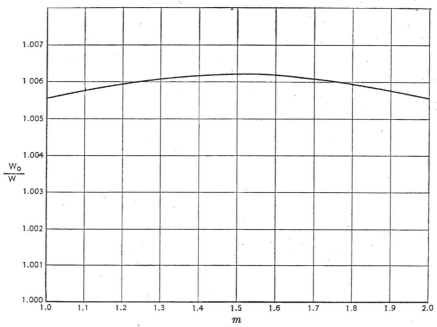

Fig. 5.—*Showing Variation of* $\dfrac{W_0}{W}$ *with m when p=0.1282.*

The case of particular interest for hysteresis is where $m=1.6$ and the corresponding values of $\dfrac{W_0}{W}$ have been worked out in Table III and platted in Fig. 6 for $p=\dfrac{1}{2}$ to $p=\dfrac{1}{19}$.

Circular Section.—The energy per cycle in an elementary ring is in this case

$$dW = \eta \left(\frac{4\pi\mu NI}{2\pi(R+x)} \right)^m 2\pi(R+x)2\sqrt{a^2-x^2}dx$$

and the total energy is

$$W = \eta\,4\pi(2\mu NI)^m \int_{-a}^{a} \frac{\sqrt{a^2-x^2}}{(R+x)^{m-1}}dx$$

Ratio of Hysteresis Loss with Uniform Distribution of Flux to the Actual Loss, Assuming Constant Permeability, in Rings of Rectangular and Circular Section.

	$\dfrac{W_o}{W}$	
	Rectangular	Circular
	m=1.6	m=1.5
$\frac{1}{2}$	1.1117	1.0841
$\frac{1}{3}$	1.0447	1.0327
$\frac{1}{4}$	1.0244	1.0183
$\frac{1}{5}$	1.0153	1.0114
$\frac{1}{6}$	1.0105	1.0077
$\frac{1}{7}$	1.0076	1.0058
$\frac{1}{8}$	1.0059	1.0041
$\frac{1}{10}$	1.0035	1.0023
$\frac{1}{19}$	1.0012	1.0008

I have not been able to work out a general solution for this integral, but it can be evaluated for particular values of m, such as 1, 2, $\frac{3}{2}$, $\frac{5}{2}$, etc.

If $m = 1$

$$W = 4\pi\eta\, 2\mu NI\, \frac{\pi a^2}{2}$$

$$= 4\pi^2 \eta\mu NIa^2$$

If $m = 2$

$$W = 4\pi\eta(2\mu NI)^2 \pi R(1 - \sqrt{1 - p^2})$$

If $m = \frac{3}{2}$

$$W = 4\pi\eta(2\mu NI)^{3/2} \int_{-a}^{a} \frac{\sqrt{a^2 - x^2}}{\sqrt{R + x}} dx.$$

This integral can be put in the form of known elliptic integrals and thus evaluated.

Let $x = a \cos 2\theta$

59629—08——8

Then

$$\int_{-a}^{a} \frac{\sqrt{a^2-x^2}}{\sqrt{R+x}}qx = \int_{0}^{\frac{\pi}{2}} \frac{2a^2\sin^2 2\theta\,d\theta}{\sqrt{R+a(1-2\sin^2\theta)}}$$

$$= \frac{8a^2}{\sqrt{R+a}} \int_{0}^{\frac{\pi}{2}} \frac{(\sin^2\theta - \sin^4\theta)d\theta}{\sqrt{1-k^2\sin^2\theta}}$$

where $k^2 = \dfrac{2a}{R+a}$. From the tables of Bierens de Haan, the values for the two integrals are

$$\int_{0}^{\frac{\pi}{2}} \frac{\sin^2\theta}{\sqrt{1-k^2\sin^2\theta}}d\theta = \frac{1}{k^2}F(k) - \frac{1}{k^2}E(k)$$

$$\int_{0}^{\frac{\pi}{2}} \frac{\sin^4\theta}{\sqrt{1-k^2\sin^2\theta}}d\theta = \frac{1}{3k^4}\Big[(2+k^2)F(k) - 2(1+k^2)E(k)\Big]$$

where $F(k)$ is the complete elliptic integral of the first kind and $E(k)$ is the complete elliptic integral of the second kind. We have, then,

$$\int_{-a}^{a} \frac{\sqrt{a^2-x^2}}{\sqrt{R+x}}dx = \frac{8a^2}{\sqrt{R+a}}\Big[\frac{2(k^2-1)}{3k^4}F(k) + \frac{2-k^2}{3k^4}E(k)\Big] = \frac{8a^2}{\sqrt{R+a}}C$$

Now

$$(R+a)^{\frac{1}{2}} = R^{\frac{1}{2}}(1+p)^{\frac{1}{2}}$$

$$= R^{\frac{1}{2}}\Big[1 + \frac{1}{2}p - \frac{1}{8}p^2 + \frac{1}{16}p^3 - \frac{5}{128}p^4 + \frac{7}{256}p^5 - \cdots\cdots\Big]$$

$$= R^{\frac{1}{2}}A$$

We have, then, for $m=\dfrac{3}{2}$

$$W = 4\pi\eta(2\mu NI)^{\frac{3}{2}}\frac{8a^2}{\sqrt{R}}\frac{C}{A}$$

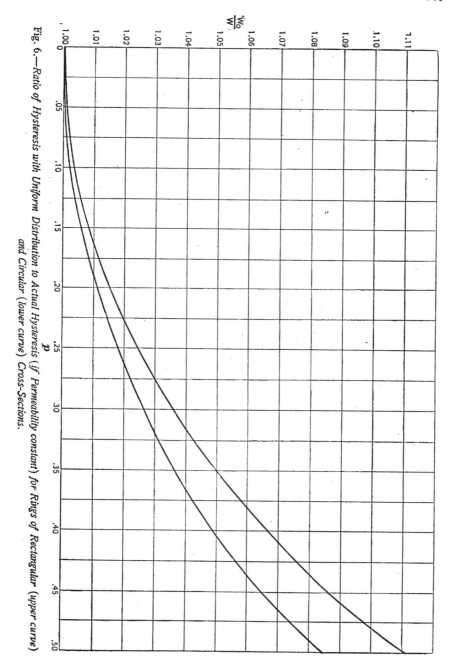

Fig. 6.—*Ratio of Hysteresis with Uniform Distribution to Actual Hysteresis (if Permeability constant) for Rings of Rectangular (upper curve) and Circular (lower curve) Cross-Sections.*

For uniform distribution

$$H_0 = \frac{4\pi NI}{\pi a^2} R(1 - \sqrt{1 - p^2})$$

$$W_0 = \eta \left[\frac{4\pi\mu NIR}{\pi a^2}(1 - \sqrt{1 - p^2}) \right]^m \int_{-a}^{a} 2\pi(R+x)2\sqrt{a^2 - x^2}\,dx$$

$$= \eta \left[\frac{4\pi\mu NIR}{\pi a^2}(1 - \sqrt{1 - p^2}) \right]^m 2\pi^2 a^2 R$$

If $m = 1$

$$W_0 = \eta \left[\frac{4\pi\mu NIR}{\pi a^2}(1 - \sqrt{1 - p^2}) \right] 2\pi^2 a^2 R$$

and

$$\frac{W_0}{W} = \frac{2}{p^2}(1 - \sqrt{1 - p^2})$$

If $m = 2$

$$W_0 = \eta(4\mu NI)^2(1 - \sqrt{1 - p^2})^2 \frac{2\pi^2 R^3}{a^2}$$

and

$$\frac{W_0}{W} = \frac{2}{p^2}(1 - \sqrt{1 - p^2})$$

If $m = \frac{3}{2}$

$$W_0 = 4\sqrt{2}\pi^2\eta(2\mu NI)^{\frac{3}{2}}\frac{R^{\frac{5}{2}}}{a}(1 - \sqrt{1 - p^2})^{\frac{3}{2}}$$

and

$$\frac{W_0}{W} = \frac{\sqrt{2}\,\pi}{8p^3}\frac{A}{C}(1 - \sqrt{1 - p^2})^{\frac{3}{2}}$$

Now

$$1-\sqrt{1-p^2}=\frac{p^2}{2}\left(1+\frac{1}{4}p^2+\frac{1}{8}p^4+\frac{5}{64}p^6+\frac{7}{128}p^8+\ \cdots\ \right)$$

$$=\frac{p^2}{2}B$$

$$(1-\sqrt{1-p^2})^{\frac{3}{2}}=\frac{p^3}{2\sqrt{2}}B^{\frac{3}{2}}$$

$$\frac{W_o}{W}=\frac{\pi}{16}\frac{A}{C}B^{\frac{3}{2}}.$$

TABLE IV.

Calculation for $m=\frac{3}{2}$ with Ring of Circular Section.

p	k²	sin⁻¹k	log E	log F	E	F	C	A	B	$\frac{\pi}{16}\frac{A}{C}B^{\frac{3}{2}}$
0.50000	0.66667	54.736	0.1007778	0.3072753	1.261182	2.028969	0.24670	1.22478	1.07178	1.0841
.33333	.50000	45.000	.1305409	.2681272	1.350644	1.854074	.22919	1.15470	1.02942	1.0327
.25000	.40000	39.232	.1459400	.2498144	1.399394	1.777520	.22085	1.11804	1.01625	1.0183
.20000	.33333	35.264	.1554326	.2390274	1.430318	1.733914	.21594	1.09545	1.01020	1.0114
.16667	.28572	32.112	.1618880	.2318830	1.451737	1.705623	.21269	1.08012	1.00704	1.0077
.14286	.25000	30.000	.1665669	.2267933	1.467462	1.685750	.21031	1.06905	1.00515	1.0058
.12500	.22222	28.125	.1701186	.2229767	1.479512	1.671000	.20864	1.06066	1.00396	1.0041
.10000	.18182	25.239	.1751520	.2176336	1.496761	1.650569	.20623	1.04881	1.00251	1.0023
.05263	.10000	18.435	.1849064	.2074841	1.530758	1.612442	.20149	1.02598	1.00069	1.0008

The values of $\frac{W_o}{W}$ for $m=1$ and $m=2$ are again equal to $\frac{H_o}{H_R}$ and may be found in Table I. For $m=\frac{3}{2}$ the values of C have been determined by the use of Legendre's Tables, while A and B are rapidly converging series and can be carried out to the necessary accuracy. The values given in Table III and platted in Fig. 6 are correct to 0.01 per cent. As the value of $\frac{W_o}{W}$ changes very slowly with m in the neighborhood of $m=\frac{3}{2}$, these values are very nearly the same as would be obtained for $m=1.6$ and represent very closely the conditions for hysteresis. The steps in the computation are shown in Table IV.

TABLE V.

Limiting Values of p and $\dfrac{1}{p}$ for Given Allowable Error in Hysteresis Measurements in Rings, Assuming Constant Permeability.

Per cent Error	Rectangular m = 1.6		Circular m = 1.5	
	p	$\dfrac{1}{p}$	p	$\dfrac{1}{p}$
0.1	0.046	22.	0.065	15.4
0.2	.070	14.3	.090	11.1
0.5	.117	8.6	.135	7.4
1.0	.162	6.2	.187	5.35
2.0	.227	4.4	.262	3.8

These results show the ratio of the loss with uniform distribution to the loss with the actual distribution, if the permeability were constant. With iron specimens the ratio may be greater or less, for with low inductions the distribution will be less uniform, while with high values of the induction it will be more uniform than with constant permeability. For the case of constant permeability, Table V gives the limiting value of p for the error to be within the assigned limit.

WASHINGTON, August 19, 1908.

O

CPSIA information can be obtained
at www.ICGtesting.com
Printed in the USA
BVHW042020300119
538843BV00013B/405/P